The Eyes Have It

by

Randall Garrett

The Eyes Have It
by Randall Garrett

ISBN: 978-93-59320-42-7

Published by

DOUBLE 9 BOOKS

2/13-B, Ansari Road
Daryaganj, New Delhi – 110002
info@double9books.com
www.double9books.com
Tel. 011-40042856

ABOUT THE AUTHOR

Phillip, Gordon Randall David Garrett (December 16, 1927 – December 31, 1987) was a science fiction and fantasy author from the United States. In the 1950s and 1960s, he contributed to Astounding and other science fiction periodicals. He taught Robert Silverberg how to market enormous amounts of action-adventure science fiction and worked with him on two novels about Earthmen upsetting a peaceful agrarian civilisation on an extraterrestrial planet. Garrett is best known for the Lord Darcy books, which include the novel Too Many Magicians and two short story collections set in an alternate world where a joint Anglo-French empire led by a Plantagenet dynasty has survived into the twentieth century and magic works and has been scientifically codified. The Darcy books are full of jokes, puns, and references (specially to works of detective and spy fiction: Lord Darcy is fashioned after Sherlock Holmes), with elements reappear frequently in the detective's lesser works. Michael Kurland went on to write two more Lord Darcy novels. Garrett used several pen names, including David Gordon, John Gordon, Darrel T. Langart (an anagram of his name), Alexander Blade, Richard Greer, Ivar Jorgensen, Clyde Mitchell, Leonard G. Spencer, S. M. Tenneshaw, and Gerald Vance. As "Randall of Hightower" (a pun on "garret"), he was also a founding member of the Society for Creative Anachronism.

Sir Pierre Morlaix, Chevalier of the Angevin Empire, Knight of the Golden Leopard, and secretary-in-private to my lord, the Count D'Evreux, pushed back the lace at his cuff for a glance at his wrist watch—three minutes of seven. The Angelus had rung at six, as always, and my lord D'Evreux had been awakened by it, as always. At least, Sir Pierre could not remember any time in the past seventeen years when my lord had not awakened at the Angelus. Once, he recalled, the sacristan had failed to ring the bell, and the Count had been furious for a week. Only the intercession of Father Bright, backed by the Bishop himself, had saved the sacristan from doing a turn in the dungeons of Castle D'Evreux.

Sir Pierre stepped out into the corridor, walked along the carpeted flagstones, and cast a practiced eye around him as he walked. These old castles were difficult to keep clean, and my lord the Count was fussy about nitre collecting in the seams between the stones of the walls. All appeared quite in order, which was a good thing. My lord the Count had been making a night of it last evening, and that always made him the more peevish in the morning. Though he always woke at the Angelus, he did not always wake up sober.

Sir Pierre stopped before a heavy, polished, carved oak door, selected a key from one of the many at his belt, and turned it in the lock. Then he went into the elevator and the door locked automatically behind him. He pressed the switch and waited in patient silence as he was lifted up four floors to the Count's personal suite.

By now, my lord the Count would have bathed, shaved, and dressed. He would also have poured down an eye-opener consisting of half a water glass of fine Champagne brandy. He would not eat breakfast until eight. The Count had no valet in the strict sense of the term. Sir Reginald Beauvay held that title, but he was never called upon to exercise the more personal functions of his office. The Count did not like to be seen until he was thoroughly presentable.

The elevator stopped. Sir Pierre stepped out into the corridor and walked along it toward the door at the far end. At exactly seven o'clock, he rapped briskly on the great door which bore the gilt-and-polychrome arms of the House D'Evreux.

For the first time in seventeen years, there was no answer.

Sir Pierre waited for the growled command to enter for a full minute, unable to believe his ears. Then, almost timidly, he rapped again.

There was still no answer.

Then, bracing himself for the verbal onslaught that would follow if he had erred, Sir Pierre turned the handle and opened the door just as if he had heard the Count's voice telling him to come in.

"Good morning, my lord," he said, as he always had for seventeen years.

But the room was empty, and there was no answer.

He looked around the huge room. The morning sunlight streamed in through the high mullioned windows and spread a diamond-checkered pattern across the tapestry on the far wall, lighting up the brilliant hunting scene in a blaze of color.

"My lord?"

Nothing. Not a sound.

The bedroom door was open. Sir Pierre walked across to it and looked in.

He saw immediately why my lord the Count had not answered, and that, indeed, he would never answer again.

My lord the Count lay flat on his back, his arms spread wide, his eyes staring at the ceiling. He was still clad in his gold and scarlet evening clothes. But the great stain on the front of his coat was not the same shade of scarlet as the rest of the cloth, and the stain had a bullet hole in its center.

Sir Pierre looked at him without moving for a long moment. Then he stepped over, knelt, and touched one of the Count's hands with the back of his own. It was quite cool. He had been dead for hours.

"I knew someone would do you in sooner or later, my lord," said Sir Pierre, almost regretfully.

Then he rose from his kneeling position and walked out without another look at his dead lord. He locked the door of the suite, pocketed the key, and went back downstairs in the elevator.

Mary, Lady Duncan stared out of the window at the morning sunlight and wondered what to do. The Angelus bell had awakened her from a fitful sleep in her chair, and she knew that, as a guest at Castle D'Evreux, she would be expected to appear at Mass again this morning. But how could she? How could she face the Sacramental Lord on the altar—to say nothing of taking the Blessed Sacrament itself.

Still, it would look all the more conspicuous if she did not show up this morning after having made it a point to attend every morning with Lady Alice during the first four days of this visit.

She turned and glanced at the locked and barred door of the bedroom. *He* would not be expected to come. Laird Duncan used his wheelchair as an excuse, but since he had taken up black magic as a hobby he had, she suspected, been actually afraid to go anywhere near a church.

If only she hadn't lied to him! But how could she have told the truth? That would have been worse—infinitely worse. And now, because of that lie, he was locked in his bedroom doing only God and the Devil knew what.

If only he would come out. If he would only stop whatever it was he had been doing for all these long hours—or at least finish it! Then they could leave Evreux, make some excuse—any excuse—to get away. One of them could feign sickness. Anything, anything to get them out of France, across the Channel, and back to Scotland, where they would be safe!

She looked back out of the window, across the courtyard, at the towering stone walls of the Great Keep and at the high window that opened into the suite of Edouard, Count D'Evreux.

Last night she had hated him, but no longer. Now there was only room in her heart for fear.

She buried her face in her hands and cursed herself for a fool. There were no tears left for weeping—not after the long night.

Behind her, she heard the sudden noise of the door being unlocked, and she turned.

Laird Duncan of Duncan opened the door and wheeled himself out. He was followed by a malodorous gust of vapor from the room he had just left. Lady Duncan stared at him.

He looked older than he had last night, more haggard and worn, and there was something in his eyes she did not like. For a moment he said nothing. Then he wet his lips with the tip of his tongue. When he spoke, his voice sounded dazed.

"There is nothing to fear any more," he said. "Nothing to fear at all."

The Reverend Father James Valois Bright, Vicar of the Chapel of Saint-Esprit, had as his flock the several hundred inhabitants of the Castle D'Evreux. As such, he was the ranking priest—socially, not hierarchically— in the country. Not counting the Bishop and the Chapter at the Cathedral, of course. But such knowledge did little good for the Father's peace of mind. The turnout of the flock was abominably small for its size—especially for week-day Masses. The Sunday Masses were well attended, of course; Count D'Evreux was there punctually at nine every Sunday, and he had a habit of counting the house. But he never showed up on weekdays, and his laxity had allowed a certain further laxity to filter down through the ranks.

The great consolation was Lady Alice D'Evreux. She was a plain, simple girl, nearly twenty years younger than her brother, the Count, and quite his opposite in every way. She was quiet where he was thundering, self-effacing where he was flamboyant, temperate where he was drunken, and chaste where he was—

Father Bright brought his thoughts to a full halt for a moment. He had, he reminded himself, no right to make judgments of that sort. He was not, after all, the Count's confessor; the Bishop was.

Besides, he should have his mind on his prayers just now.

He paused and was rather surprised to notice that he had already put on his alb, amice, and girdle, and he was aware that his lips had formed the words of the prayer as he had donned each of them.

Habit, he thought, *can be destructive to the contemplative faculty.*

He glanced around the sacristy. His server, the young son of the Count of Saint Brieuc, sent here to complete his education as a gentleman who would some day be the King's Governor of one of the most important counties in Brittany, was pulling his surplice down over his head. The clock said 7:11.

Father Bright forced his mind Heavenward and repeated silently the vesting prayers that his lips had formed meaninglessly, this time putting his full intentions behind them. Then he added a short mental prayer asking God to forgive him for allowing his thoughts to stray in such a manner.

He opened his eyes and reached for his chasuble just as the sacristy door opened and Sir Pierre, the Count's Privy Secretary, stepped in.

"I must speak to you, Father," he said in a low voice. And, glancing at the young De Saint-Brieuc, he added: "Alone."

Normally, Father Bright would have reprimanded anyone who presumed to break into the sacristy as he was vesting for Mass, but he knew that Sir Pierre would never interrupt without good reason. He nodded and went outside in the corridor that led to the altar.

"What is it, Pierre?" he asked.

"My lord the Count is dead. Murdered."

After the first momentary shock, Father Bright realized that the news was not, after all, totally unexpected. Somewhere in the back of his mind, it seemed he had always known that the Count would die by violence long before debauchery ruined his health.

"Tell me about it," he said quietly.

Sir Pierre reported exactly what he had done and what he had seen.

"Then I locked the door and came straight here," he told the priest.

"Who else has the key to the Count's suite?" Father Bright asked.

"No one but my lord himself," Sir Pierre answered, "at least as far as I know."

"Where is his key?"

"Still in the ring at his belt. I noticed that particularly."

"Very good. We'll leave it locked. You're certain the body was cold?"

"Cold and waxy, Father."

"Then he's been dead many hours."

"Lady Alice will have to be told," Sir Pierre said.

Father Bright nodded. "Yes. The Countess D'Evreux must be informed of her succession to the County Seat." He could tell by the sudden momentary blank look that came over Sir Pierre's face that the Privy Secretary had not yet realized fully the implications of the Count's death. "I'll tell her, Pierre. She should be in her pew by now. Just step into the church and tell her quietly that I want to speak to her. Don't tell her anything else."

"I understand, Father," said Sir Pierre.

There were only twenty-five or thirty people in the pews—most of them women—but Alice, Countess D'Evreux was not one of them. Sir Pierre walked quietly and unobtrusively down the side aisle and out into the narthex. She was standing there, just inside the main door, adjusting the black lace mantilla about her head, as though she had just come in from outside. Suddenly, Sir Pierre was very glad he would not have to be the one to break the news.

She looked rather sad, as always, her plain face unsmiling. The jutting nose and square chin which had given her brother the Count a look of aggressive handsomeness only made her look very solemn and rather sexless, although she had a magnificent figure.

"My lady," Sir Pierre said, stepping towards her, "the Reverent Father would like to speak to you before Mass. He's waiting at the sacristy door."

She held her rosary clutched tightly to her breast and gasped. Then she said, "Oh. Sir Pierre. I'm sorry; you quite surprised me. I didn't see you."

"My apologies, my lady."

"It's all right. My thoughts were elsewhere. Will you take me to the good Father?"

Father Bright heard their footsteps coming down the corridor before he saw them. He was a little fidgety because Mass was already a minute overdue. It should have started promptly at 7:15.

The new Countess D'Evreux took the news calmly, as he had known she would. After a pause, she crossed herself and said: "May his soul rest in peace. I will leave everything in your hands, Father, Sir Pierre. What are we to do?"

"Pierre must get on the teleson to Rouen immediately and report the matter to His Highness. I will announce your brother's death and ask for prayers for his soul—but I think I need say nothing about the manner of his death. There is no need to arouse any more speculation and fuss than necessary."

"Very well," said the Countess. "Come, Sir Pierre; I will speak to the Duke, my cousin, myself."

"Yes, my lady."

Father Bright returned to the sacristy, opened the missal, and changed the placement of the ribbons. Today was an ordinary Feria; a Votive Mass would not be forbidden by the rubics. The clock said 7:17. He turned to young De Saint-Brieuc, who was waiting respectfully. "Quickly, my son— go and get the unbleached beeswax candles and put them on the altar. Be sure you light them before you put out the white ones. Hurry, now; I will be ready by the time you come back. Oh yes—and change the altar frontal. Put on the black."

"Yes, Father." And the lad was gone.

Father Bright folded the green chasuble and returned it to the drawer, then took out the black one. He would say a Requiem for the Souls of All the Faithful Departed—and hope that the Count was among them.

His Royal Highness, the Duke of Normandy, looked over the official letter his secretary had just typed for him. It was addressed to *Serenissimus Dominus Nostrus Iohannes Quartus, Dei Gratia, Angliae, Franciae, Scotiae, Hiberniae, et Novae Angliae, Rex, Imperator, Fidei Defensor,* ... "Our Most Serene Lord, John IV, by the Grace of God King and Emperor of England, France, Scotland, Ireland, and New England, Defender of the Faith, ..."

It was a routine matter; simple notification to his brother, the King, that His Majesty's most faithful servant, Edouard, Count of Evreux had departed this life, and asking His Majesty's confirmation of the Count's heir-at-law, Alice, Countess of Evreux as his lawful successor.

His Highness finished reading, nodded, and scrawled his signature at the bottom: *Richard Dux Normaniae.*

Then, on a separate piece of paper, he wrote: "Dear John, May I suggest you hold up on this for a while? Edouard was a lecher and a slob, and I have no doubt he got everything he deserved, but we have no notion who killed him. For any evidence I have to the contrary, it might have been Alice who pulled the trigger. I will send you full particulars as soon as I have them. With much love, Your brother and servant, Richard."

He put both papers into a prepared envelope and sealed it. He wished he could have called the king on the teleson, but no one had yet figured out how to get the wires across the channel.

He looked absently at the sealed envelope, his handsome blond features thoughtful. The House of Plantagenet had endured for eight centuries, and the blood of Henry of Anjou ran thin in its veins, but the Norman strain was as strong as ever, having been replenished over the centuries by fresh infusions from Norwegian and Danish princesses. Richard's mother, Queen Helga, wife to His late Majesty, Henry X, spoke very few words of Anglo-French, and those with a heavy Norse accent.

Nevertheless, there was nothing Scandinavian in the language, manner, or bearing of Richard, Duke of Normandy. Not only was he a member of the oldest and most powerful ruling family of Europe, but he bore a Christian name that was distinguished even in that family. Seven Kings of the Empire had borne the name, and most of them had been good Kings—if not always "good" men in the nicey-nicey sense of the word. Even old Richard I, who'd been pretty wild during the first forty-odd years of his life, had settled down to do a magnificent job of kinging for the next twenty years. The long and painful recovery from the wound he'd received at the Siege of Chaluz had made a change in him for the better.

There was a chance that Duke Richard might be called upon to uphold the honor of that name as King. By law, Parliament must elect a Plantagenet as King in the event of the death of the present Sovereign, and while the election of one of the King's two sons, the Prince of Wales and the Duke of Lancaster, was more likely than the election of Richard, he was certainly not eliminated from the succession.

Meantime, he would uphold the honor of his name as Duke of Normandy.

Murder had been done; therefore justice must be done. The Count D'Evreux had been known for his stern but fair justice almost as well as he had been known for his profligacy. And, just as his pleasures had been without temperance, so his justice had been untempered by mercy. Whoever had killed him would find both justice and mercy—in so far as Richard had it within his power to give it.

Although he did not formulate it in so many words, even mentally, Richard was of the opinion that some debauched woman or cuckolded man had fired the fatal shot. Thus he found himself inclining toward mercy before he knew anything substantial about the case at all.

Richard dropped the letter he was holding into the special mail pouch that would be placed aboard the evening trans-Channel packet, and then turned in his chair to look at the lean, middle-aged man working at a desk across the room.

"My lord Marquis," he said thoughtfully.

"Yes, Your Highness?" said the Marquis of Rouen, looking up.

"How true are the stories one has heard about the late Count?"

"True, Your Highness?" the Marquis said thoughtfully. "I would hesitate to make any estimate of percentages. Once a man gets a reputation like that, the number of his reputed sins quickly surpasses the number of actual ones. Doubtless many of the stories one hears are of whole cloth; others may have only a slight basis in fact. On the other hand, it is highly likely that there are many of which we have never heard. It is absolutely certain, however, that he has acknowledged seven illegitimate sons, and I dare say he has ignored a few daughters—and these, mind you, with unmarried women. His adulteries would be rather more difficult to establish, but I think your Highness can take it for granted that such escapades were far from uncommon."

He cleared his throat and then added, "If Your Highness is looking for motive, I fear there is a superabundance of persons with motive."

"I see," the Duke said. "Well, we will wait and see what sort of information Lord Darcy comes up with." He looked up at the clock. "They should be there by now."

Then, as if brushing further thoughts on the subject from his mind, he went back to work, picking up a new sheaf of state papers from his desk.

The Marquis watched him for a moment and smiled a little to himself. The young Duke took his work seriously, but was well-balanced about it. A little inclined to be romantic—but aren't we all at nineteen? There was no doubt of his ability, nor of his nobility. The Royal Blood of England always came through.

"My lady," said Sir Pierre gently, "the Duke's Investigators have arrived."

My Lady Alice, Countess D'Evreux, was seated in a gold-brocade upholstered chair in the small receiving room off the Great Hall. Standing near her, looking very grave, was Father Bright. Against the blaze of color on the walls of the room, the two of them stood out like ink blots. Father Bright wore his normal clerical black, unrelieved except for the pure white lace at collar and cuffs. The Countess wore unadorned black velvet, a dress which she had had to have altered hurriedly by her dressmaker; she had always hated black and owned only the mourning she had worn when her mother died eight years before. The somber looks on their faces seemed to make the black blacker.

"Show them in, Sir Pierre," the Countess said calmly.

Sir Pierre opened the door wider, and three men entered. One was dressed as one gently born; the other two wore the livery of the Duke of Normandy.

The gentleman bowed. "I am Lord Darcy, Chief Criminal Investigator for His Highness, the Duke, and your servant, my lady." He was a tall, brown-haired man in his thirties with a rather handsome, lean face. He spoke Anglo-French with a definite English accent.

"My pleasure, Lord Darcy," said the Countess. "This is our vicar, Father Bright."

"Your servant, Reverend Sir." Then he presented the two men with him. The first was a scholarly-looking, graying man wearing pince-nez glasses with gold rims, Dr. Pateley, Physician. The second, a tubby, red-faced, smiling man, was Master Sean O Lochlainn, Sorcerer.

As soon as Master Sean was presented he removed a small, leather-bound folder from his belt pouch and proffered it to the priest. "My license, Reverend Father."

Father Bright took it and glanced over it. It was the usual thing, signed and sealed by the Archbishop of Rouen. The law was rather strict on that point; no sorcerer could practice without the permission of the Church, and a license was given only after careful examination for orthodoxy of practice.

"It seems to be quite in order, Master Sean," said the priest, handing the folder back. The tubby little sorcerer bowed his thanks and returned the folder to his belt pouch.

Lord Darcy had a notebook in his hand. "Now, unpleasant as it may be, we shall have to check on a few facts." He consulted his notes, then looked up at Sir Pierre. "You, I believe, discovered the body?"

"That is correct, your lordship."

"How long ago was this?"

Sir Pierre glanced at his wrist watch. It was 9:55. "Not quite three hours ago, your lordship."

"At what time, precisely?"

"I rapped on the door precisely at seven, and went in a minute or two later—say 7:01 or 7:02."

"How do you know the time so exactly?"

"My lord the Count," said Sir Pierre with some stiffness, "insisted upon exact punctuality. I have formed the habit of referring to my watch regularly."

"I see. Very good. Now, what did you do then?"

Sir Pierre described his actions briefly.

"The door to his suite was not locked, then?" Lord Darcy asked.

"No, sir."

"You did not expect it to be locked?"

"No, sir. It has not been for seventeen years."

Lord Darcy raised one eyebrow in a polite query. "Never?"

"Not at seven o'clock, your lordship. My lord the Count always rose promptly at six and unlocked the door before seven."

"He did lock it at night, then?"

"Yes, sir."

Lord Darcy looked thoughtful and made a note, but he said nothing more on that subject. "When you left, you locked the door?"

"That is correct, your lordship."

"And it has remained locked ever since?"

Sir Pierce hesitated and glanced at Father Bright. The priest said: "At 8:15, Sir Pierre and I went in. I wished to view the body. We touched nothing. We left at 8:20."

Master Sean O Lochlainn looked agitated. "Er ... excuse me, Reverend Sir. You didn't give him Holy Unction, I hope?"

"No," said Father Bright. "I thought it would be better to delay that until after the authorities has seen the ... er ... scene of the crime. I wouldn't want to make the gathering of evidence any more difficult than necessary."

"Quite right," murmured Lord Darcy.

"No blessings, I trust, Reverend Sir?" Master Sean persisted. "No exorcisms or—"

"Nothing," Father Bright interrupted somewhat testily. "I believe I crossed myself when I saw the body, but nothing more."

"Crossed *yourself*, sir. Nothing else?"

"No."

"Well, that's all right, then. Sorry to be so persistent, Reverend Sir, but any miasma of evil that may be left around is a very important clue, and it shouldn't be dispersed until it's been checked, you see."

"*Evil?*" My lady the Countess looked shocked.

"Sorry, my lady, but—" Master Sean began contritely.

But Father Bright interrupted by speaking to the Countess. "Don't distress yourself, my daughter; these men are only doing their duty."

"Of course. I understand. It's just that it's so—" She shuddered delicately.

Lord Darcy cast Master Sean a warning look, then asked politely, "Has my lady seen the deceased?"

"No," she said. "I will, however, if you wish."

"We'll see," said Lord Darcy. "Perhaps it won't be necessary. May we go up to the suite now?"

"Certainly," the Countess said. "Sir Pierre, if you will?"

"Yes, my lady."

As Sir Pierre unlocked the emblazoned door, Lord Darcy said: "Who else sleeps on this floor?"

"No one else, your lordship," Sir Pierre said. "The entire floor is ... was ... reserved for my lord the Count."

"Is there any way up besides that elevator?"

Sir Pierre turned and pointed toward the other end of the short hallway. "That leads to the staircase," he said, pointing to a massive oaken door, "but it's kept locked at all times. And, as you can see, there is a heavy bar across it. Except for moving furniture in and out or something like that, it's never used."

"No other way up or down, then?"

Sir Pierre hesitated. "Well, yes, your lordship, there is. I'll show you."

"A secret stairway?"

"Yes, your lordship."

"Very well. We'll look at it after we've seen the body."

Lord Darcy, having spent an hour on the train down from Rouen, was anxious to see the cause of the trouble at last.

He lay in the bedroom, just as Sir Pierre and Father Bright had left him.

"If you please, Dr. Pateley," said his lordship.

He knelt on one side of the corpse and watched carefully while Pateley knelt on the other side and looked at the face of the dead man. Then he touched one of the hands and tried to move an arm. "Rigor has set in— even to the fingers. Single bullet hole. Rather small caliber—I should say a .28 or .34—hard to tell until I've probed out the bullet. Looks like it went right through the heart, though. Hard to tell about powder burns; the blood has soaked the clothing and dried. Still, these specks ... hm-m-m. Yes. Hm-m-m."

Lord Darcy's eyes took in everything, but there was little enough to see on the body itself. Then his eye was caught by something that gave off a golden gleam. He stood up and walked over to the great canopied four-poster bed, then he was on his knees again, peering under it. A coin? No.

He picked it up carefully and looked at it. A button. Gold, intricately engraved in an Arabesque pattern, and set in the center with a single diamond. How long had it lain there? Where had it come from? Not from the Count's clothing, for his buttons were smaller, engraved with his arms, and had no gems. Had a man or a woman dropped it? There was no way of knowing at this stage of the game.

Darcy turned to Sir Pierre. "When was this room last cleaned?"

"Last evening, your lordship," the secretary said promptly. "My lord was always particular about that. The suite was always to be swept and cleaned during the dinner hour."

"Then this must have rolled under the bed at some time after dinner. Do you recognize it? The design is distinctive."

The Privy Secretary looked carefully at the button in the palm of Lord Darcy's hand without touching it. "I ... I hesitate to say," he said at last. "It looks like ... but I'm not sure—"

"Come, come, Chevalier! Where do you think you *might* have seen it? Or one like it." There was a sharpness in the tone of his voice.

"I'm not trying to conceal anything, your lordship," Sir Pierre said with equal sharpness. "I said I was not sure. I still am not, but it can be checked easily enough. If your lordship will permit me—" He turned and spoke to Dr. Pateley, who was still kneeling by the body. "May I have my lord the Count's keys, doctor?"

Pateley glanced up at Lord Darcy, who nodded silently. The physician detached the keys from the belt and handed them to Sir Pierre.

The Privy Secretary looked at them for a moment, then selected a small gold key. "This is it," he said, separating it from the others on the ring. "Come with me, your lordship."

Darcy followed him across the room to a broad wall covered with a great tapestry that must have dated back to the sixteenth century. Sir Pierre reached behind it and pulled a cord. The entire tapestry slid aside like a panel, and Lord Darcy saw that it was supported on a track some ten feet from the floor. Behind it was what looked at first like ordinary oak paneling, but Sir Pierre fitted the small key into an inconspicuous hole and turned. Or, rather, tried to turn.

"That's odd," said Sir Pierre. "It's not locked!"

He took the key out and pressed on the panel, shoving sideways with his hand to move it aside. It slid open to reveal a closet.

The closet was filled with women's clothing of all kinds, and styles.

Lord Darcy whistled soundlessly.

"Try that blue robe, your lordship," the Privy Secretary said. "The one with the—Yes, that's the one."

Lord Darcy took it off its hanger. The same buttons. They matched. And there was one missing from the front! Torn off! "Master Sean!" he called without turning.

Master Sean came with a rolling walk. He was holding an oddly-shaped bronze thing in his hand that Sir Pierre didn't quite recognize. The sorcerer was muttering. "Evil, that there is! Faith, and the vibrations are all over the place. Yes, my lord?"

"Check this dress and the button when you get round to it. I want to know when the two parted company."

"Yes, my lord." He draped the robe over one arm and dropped the button into a pouch at his belt. "I can tell you one thing, my lord. You talk about an evil miasma, this room has got it!" He held up the object in his hand. "There's an underlying background—something that has been here for years, just seeping in. But on top of that, there's a hellish big blast of it superimposed. Fresh it is, and very strong."

"I shouldn't be surprised, considering there was murder done here last night—or very early this morning," said Lord Darcy.

"Hm-m-m, yes. Yes, my lord, the death is there—but there's something else. Something I can't place."

"You can tell that just by holding that bronze cross in your hand?" Sir Pierre asked interestedly.

Master Sean gave him a friendly scowl. "'Tisn't quite a cross, sir. This is what is known as a *crux ansata*. The ancient Egyptians called it an *ankh*. Notice the loop at the top instead of the straight piece your true cross has.

Now, your true cross—if it were properly energized, blessed, d'ye see— your true cross would tend to dissipate the evil. The *ankh* merely vibrates to evil because of the closed loop at the top, which makes a return circuit. And it's not energized by blessing, but by another ... um ... spell."

"Master Sean, we have a murder to investigate," said Lord Darcy.

The sorcerer caught the tone of his voice and nodded quickly. "Yes, my lord." And he walked rollingly away.

"Now where's that secret stairway you mentioned, Sir Pierre?" Lord Darcy asked.

"This way, your lordship."

He led Lord Dacy to a wall at right angles to the outer wall and slid back another tapestry.

"Good Heavens," Darcy muttered, "does he have something concealed behind every arras in the place?" But he didn't say it loud enough for the Privy Secretary to hear.

This time, what greeted them was a solid-seeming stone wall. But Sir Pierre pressed in on one small stone, and a section of the wall swung back, exposing a stairway.

"Oh, yes," Darcy said. "I see what he did. This is the old spiral stairway that goes round the inside of the Keep. There are two doorways at the bottom. One opens into the courtyard, the other is a postern gate through the curtain wall to the outside—but that was closed up in the sixteenth century, so the only way out is into the courtyard."

"Your lordship knows Castle D'Evreux, then?" Sir Pierre said. The knight himself was nearly fifty, while Darcy was only in his thirties, and Sir Pierre had no recollection of Darcy's having been in the castle before.

"Only by the plans in the Royal Archives. But I have made it a point to—" He stopped. "Dear me," he interrupted himself mildly, "what is that?"

"That" was something that had been hidden by the arras until Sir Pierre had slid it aside, and was still showing only a part of itself. It lay on the floor a foot or so from the secret door.

Darcy knelt down and pulled the tapestry back from the object. "Well, well. A .28 two-shot pocket gun. Gold-chased, beautifully engraved, mother-of-pearl handle. A regular gem." He picked it up and examined it closely. "One shot fired."

He stood up and showed it to Sir Pierre. "Ever see it before?"

The Privy Secretary looked at the weapon closely. Then he shook his head. "Not that I recall, your lordship. It certainly isn't one of the Count's guns."

"You're certain?"

"Quite certain, your lordship. I'll show you the gun collection if you want. My lord the Count didn't like tiny guns like that; he preferred a larger caliber. He would never have owned what he considered a toy."

"Well, we'll have to look into it." He called over Master Sean again and gave the gun into his keeping. "And keep your eyes open for anything else of interest, Master Sean. So far, everything of interest besides the late Count himself has been hiding under beds or behind arrases. Check everything. Sir Pierre and I are going for a look down this stairway."

The stairway was gloomy, but enough light came in through the arrow slits spaced at intervals along the outer way to illuminate the interior. It spiraled down between the inner and outer walls of the Great Keep, making four complete circuits before it reached ground level. Lord Darcy looked carefully at the steps, the walls, and even the low, arched overhead as he and Sir Pierre went down.

After the first circuit, on the floor beneath the Count's suite, he stopped. "There was a door here," he said, pointing to a rectangular area in the inner wall.

"Yes, your lordship. There used to be an opening at every floor, but they were all sealed off. It's quite solid, as you can see."

"Where would they lead if they were open?"

"The county offices. My own office, the clerk's offices, the constabulary on the first floor. Below are the dungeons. My lord the Count was the only one who lived in the Keep itself. The rest of the household live above the Great Hall."

"What about guests?"

"They're usually housed in the east wing. We only have two house guests at the moment. Laird and Lady Duncan have been with us for four days."

"I see." They went down perhaps four more steps before Lord Darcy asked quietly, "Tell me, Sir Pierre, were you privy to *all* of Count D'Evreux's business?"

Another four steps down before Sir Pierre answered. "I understand what your lordship means," he said. Another two steps. "No, I was not. I was aware that my lord the Count engaged in certain ... er ... shall we say, liaisons with members of the opposite sex. However—"

He paused, and in the gloom, Lord Darcy could see his lips tighten. "However," he continued, "I did not procure for my lord, if that is what you're driving at. I am not and never have been a pimp."

"I didn't intend to suggest that you had, good knight," said Lord Darcy in a tone that strongly implied that the thought had actually never crossed his mind. "Not at all. But certainly there is a difference between 'aiding and abetting' and simple knowledge of what is going on."

"Oh. Yes. Yes, of course. Well, one cannot, of course, be the secretary-in-private of a gentleman such as my lord the Count for seventeen years without knowing something of what is going on, you're right. Yes. Yes. Hm-m-m."

Lord Darcy smiled to himself. Not until this moment had Sir Pierre realized how much he actually *did* know. In loyalty to his lord, he had literally kept his eyes shut for seventeen years.

"I realize," Lord Darcy said smoothly, "that a gentleman would never implicate a lady nor besmirch the reputation of another gentleman without due cause and careful consideration. However,"—like the knight, he paused a moment before going on—"although we are aware that he was not discreet, was he particular?"

"If you mean by that, did he confine his attentions to those of gentle birth, your lordship, then I can say, no he did not. If you mean did he confine his attentions to the gentler sex, then I can only say that, as far as I know, he did."

"I see. That explains the closet full of clothes."

"Beg pardon, your lordship?"

"I mean that if a girl or woman of the lower classes were to come here, he would have proper clothing for them to wear—in spite of the sumptuary laws to the contrary."

"Quite likely, your lordship. He was most particular about clothing. Couldn't stand a woman who was sloppily dressed or poorly dressed."

"In what way?"

"Well. Well, for instance, I recall once that he saw a very pretty peasant girl. She was dressed in the common style, of course, but she was dressed neatly and prettily. My lord took a fancy to her. He said, 'Now there's a lass who knows how to wear clothes. Put her in decent apparel, and she'd pass for a princess.' But a girl, who had a pretty face and a fine figure, made no impression on him unless she wore her clothing well, if you see what I mean, your lordship."

"Did you ever know him to fancy a girl who dressed in an offhand manner?" Lord Darcy asked.

"Only among the gently born, your lordship. He'd say, 'Look at Lady So-and-so! Nice wench, if she'd let me teach her how to dress.' You might say, your lordship, that a woman could be dressed commonly or sloppily, but not both."

"Judging by the stuff in that closet," Lord Darcy said, "I should say that the late Count had excellent taste in feminine dress."

Sir Pierre considered. "Hm-m-m. Well, now, I wouldn't exactly say so, your lordship. He knew *how* clothes should be worn, yes. But he couldn't pick out a woman's gown of his own accord. He could choose his own clothing with impeccable taste, but he'd not any real notion of how a woman's clothing should go, if you see what I mean. All he knew was how good clothing should be worn. But he knew nothing about design for women's clothing."

"Then how did he get that closet full of clothes?" Lord Darcy asked, puzzled.

Sir Pierre chuckled. "Very simply, your lordship. He knew that the Lady Alice had good taste, so he secretly instructed that each piece that Lady Alice ordered should be made in duplicate. With small variations, of course. I'm certain my lady wouldn't like it if she knew."

"I dare say not," said Lord Darcy thoughtfully.

"Here is the door to the courtyard," said Sir Pierre. "I doubt that it has been opened in broad daylight for many years." He selected a key from the ring of the late Count and inserted it into the keyhole. The door swung back, revealing a large crucifix attached to its outer surface. Lord Darcy crossed himself. "Lord in Heaven," he said softly, "what is this?"

He looked out into a small shrine. It was walled off from the courtyard and had a single small entrance some ten feet from the doorway. There were four *prie-dieus*—small kneeling benches—ranged in front of the doorway.

"If I may explain, your lordship—" Sir Pierre began.

"No need to," Lord Darcy said in a hard voice. "It's rather obvious. My lord the Count was quite ingenious. This is a relatively newly-built shrine. Four walls and a crucifix against the castle wall. Anyone could come in here, day or night, for prayer. No one who came in would be suspected." He stepped out into the small enclosure and swung around to look at the door. "And when that door is closed, there is no sign that there is a door behind the crucifix. If a woman came in here, it would be assumed that she came for prayer. But if she knew of that door—" His voice trailed off.

"Yes, your lordship," said Sir Pierre. "I did not approve, but I was in no position to disapprove."

"I understand." Lord Darcy stepped out to the doorway of the little shrine and took a quick glance about. "Then anyone within the castle walls could come in here," he said.

"Yes, your lordship."

"Very well. Let's go back up."

In the small office which Lord Darcy and his staff had been assigned while conducting the investigation, three men watched while a fourth conducted a demonstration on a table in the center of the room.

Master Sean O Lochlainn held up an intricately engraved gold button with an Arabesque pattern and a diamond set in the center.

He looked at the other three. "Now, my lord, your Reverence, and colleague Doctor, I call your attention to this button."

Dr. Pateley smiled and Father Bright looked stern. Lord Darcy merely stuffed tobacco—imported from the southern New England counties on the Gulf—into a German-made porcelain pipe. He allowed Master Sean a certain amount of flamboyance; good sorcerers were hard to come by.

"Will you hold the robe, Dr. Pateley? Thank you. Now, stand back. That's it. Thank you. Now, I place the button on the table, a good ten feet from the robe." Then he muttered something under his breath and dusted a bit of powder on the button. He made a few passes over it with his hands, paused, and looked up at Father Bright. "If you will, Reverend Sir?"

Father Bright solemnly raised his right hand, and, as he made the Sign of the Cross, said: "May this demonstration, O God, be in strict accord with the truth, and may the Evil One not in any way deceive us who are witnesses thereto. In the Name of the Father and of the Son and of the Holy Spirit. Amen."

"Amen," the other three chorused.

Master Sean crossed himself, then muttered something under his breath.

The button leaped from the table, slammed itself against the robe which Dr. Pateley held before him, and stuck there as though it had been sewed on by an expert.

"Ha!" said Master Sean. "As I thought!" He gave the other three men a broad, beaming smile. "The two were definitely connected!"

Lord Darcy looked bored. "Time?" he asked.

"In a moment, my lord," Master Sean said apologetically. "In a moment." While the other three watched, the sorcerer went through more spells with the button and the robe, although none were quite so spectacular as the first demonstration. Finally, Master Sean said: "About eleven thirty last night they were torn apart, my lord. But I shouldn't like to make it any more definite than to say between eleven and midnight. The speed with which it returned to its place shows that it was ripped off very rapidly, however."

"Very good," said Lord Darcy. "Now the bullet, if you please."

"Yes, my lord. This will have to be a bit different." He took more paraphernalia out of his large, symbol-decorated carpet bag. "The Law of Contagion, gently-born sirs, is a tricky thing to work with. If a man doesn't know how to handle it, he can get himself killed. We had an apprentice o' the guild back in Cork who might have made a good sorcerer in time. He had the talent—unfortunately, he didn't have the good sense to go with it. According to the Law of Contagion any two objects which have ever been in contact with each other have an affinity for each other which is directly proportional to the product of the degree of relevancy of the contact and the length of time they were in contact and inversely proportional to the length of time since they have ceased to be in contact." He gave a smiling glance to the priest. "That doesn't apply strictly to relics of the saints, Reverend Sir; there's another factor enters in there, as you know."

As he spoke, the sorcerer was carefully clamping the little handgun into the padded vise so that its barrel was parallel to the surface of the table.

"Anyhow," he went on, "this apprentice, all on his own, decided to get rid of the cockroaches in his house—a simple thing, if one knows how to go about it. So he collected dust from various cracks and crannies about the house, dust which contained, of course, the droppings of the pests. The dust, with the appropriate spells and ingredients, he boiled. It worked fine.

The roaches all came down with a raging fever and died. Unfortunately, the clumsy lad had poor laboratory technique. He allowed three drops of his own perspiration to fall into the steaming pot over which he was working, and the resulting fever killed him, too."

By this time, he had put the bullet which Dr. Pateley had removed from the Count's body on a small pedestal so that it was exactly in line with the muzzle of the gun. "There now," he said softly.

Then he repeated the incantation, and the powdering that he had used on the button. As the last syllable was formed by his lips, the bullet vanished with a *ping*! In its vise, the little gun vibrated.

"Ah!" said Master Sean. "No question there, eh? That's the death weapon, all right, my lord. Yes. Time's almost exactly the same as that of the removal of the button. Not more than a few seconds later. Forms a picture, don't it, my lord? His lordship the Count jerks a button off the girl's gown, she outs with a gun and plugs him."

Lord Darcy's handsome face scowled. "Let's not jump to any hasty conclusions, my good Sean. There is no evidence whatever that he was killed by a woman."

"Would a man be wearing that gown, my lord?"

"Possibly," said Lord Darcy. "But who says that anyone was wearing it when the button was removed?"

"Oh." Master Sean subsided into silence. Using a small ramrod, he forced the bullet out of the chamber of the little pistol.

"Father Bright," said Lord Darcy, "will the Countess be serving tea this afternoon?"

The priest looked suddenly contrite. "Good heavens! None of you has eaten yet! I'll see that something is sent up right away, Lord Darcy. In the confusion—"

Lord Darcy held up a hand. "I beg your pardon, Father; that wasn't what I meant. I'm sure Master Sean and Dr. Pateley would appreciate a little something, but I can wait until tea time. What I was thinking was that perhaps the Countess would ask her guests to tea. Does she know Laird and Lady Duncan well enough to ask for their sympathetic presence on such an afternoon as this?"

Father Bright's eyes narrowed a trifle. "I dare say it could be arranged, Lord Darcy. You will be there?"

"Yes—but I may be a trifle late. That will hardly matter at an informal tea."

The priest glanced at his watch. "Four o'clock?"

"I should think that would do it," said Lord Darcy.

Father Bright nodded wordlessly and left the room.

Dr. Pateley took off his pince-nez and polished the lenses carefully with a silk handkerchief. "How long will your spell keep the body incorrupt, Master Sean?" he asked.

"As long as it's relevant. As soon as the case is solved, or we have enough data to solve the case—as the case may be, heh heh—he'll start to go. I'm not a saint, you know; it takes powerful motivation to keep a body incorrupt for years and years."

Sir Pierre was eying the gown that Pateley had put on the table. The button was still in place, as if held there by magnetism. He didn't touch it. "Master Sean, I don't know much about magic," he said, "but can't you find out who was wearing this robe just as easily as you found out that the button matched?"

Master Sean wagged his head in a firm negative. "No, sir. 'Tisn't relevant sir. The relevancy of the integrated dress-as-a-whole is quite strong. So is that of the seamstress or tailor who made the garment, and that of the weaver who made the cloth. But, except in certain circumstances, the person who wears or wore the garment has little actual relevancy to the garment itself."

"I'm afraid I don't understand," said Sir Pierre, looking puzzled.

"Look at it like this, sir: That gown wouldn't be what it is if the weaver hadn't made the cloth in that particular way. It wouldn't be what it is if the seamstress hadn't cut it in a particular way and sewed it in a specific manner. You follow, sir? Yes. Well, then, the connections between garment-and-weaver and garment-and-seamstress are strongly relevant. But this dress would still be pretty much what it is if it had stayed in the closet instead of being worn. No relevance—or very little. Now, if it were a well-worn garment, that would be different—that is, if it had always been worn by the same person. Then, you see, sir, the garment-as-a-whole is what it is because of the wearing, and the wearer becomes relevant."

He pointed at the little handgun he was still holding in his hand. "Now you take your gun, here, sir. The—"

"It isn't my gun," Sir Pierre interrupted firmly.

"I was speaking rhetorically, sir," said Master Sean with infinite patience. "This gun or any other gun in general, if you see what I mean, sir. It's even harder to place the ownership of a gun. Most of the wear on a gun is purely mechanical. It don't matter *who* pulls the trigger, you see, the erosion by the gases produced in the chamber, and the wear caused by the bullet passing through the barrel will be the same. You see, sir, 'tisn't relevant *to the gun* who pulled its trigger or what it's fired at. The bullet's a slightly different matter. To the bullet, it *is* relevant which gun it was fired from and what it hit. All these things simply have to be taken into account, Sir Pierre."

"I see," said the knight. "Very interesting, Master Sean." Then he turned to Lord Darcy. "Is there anything else, your lordship? There's a great deal of county business to be attended to."

Lord Darcy waved a hand. "Not at the moment, Sir Pierre. I understand the pressures of government. Go right ahead."

"Thank you, your lordship. If anything further should be required, I shall be in my office."

As soon as Sir Pierre had closed the door, Lord Darcy held out his hand toward the sorcerer. "Master Sean; the gun."

Master Sean handed it to him. "Ever see one like it before?" he asked, turning it over in his hands.

"Not *exactly* like it, my lord."

"Come, come, Sean; don't be so cautious. I am no sorcerer, but I don't need to know the Laws of Similarity to be able to recognise an *obvious* similarity."

"Edinburgh," said Master Sean flatly.

"Exactly. Scottish work. The typical Scot gold work; remarkable beauty. And look at that lock. It has 'Scots' written all over it—and more. 'Edinburgh', as you said."

Dr. Pateley, having replaced his carefully polished glasses, leaned over and peered at the weapon in Lord Darcy's hand. "Couldn't it be Italian, my lord? Or Moorish? In Moorish Spain, they do work like that."

"No Moorish gunsmith would put a hunting scene on the butt," Lord Darcy said flatly, "and the Italians wouldn't have put heather and thistles in the field surrounding the huntsman."

"But the *FdM* engraved on the barrel," said Dr. Pateley, "indicates the—"

"Ferrari of Milan," said Lord Darcy. "Exactly. But the barrel is of much newer work than the rest. So are the chambers. This is a fairly old gun—fifty years old, I'd say. The lock and the butt are still in excellent condition, indicating that it has been well cared for, but frequent usage—or a single accident—could ruin the barrel and require the owner to get a replacement. It was replaced by Ferrari."

"I see," said Dr. Pateley somewhat humbled.

"If we open the lock ... Master Sean, hand me your small screwdriver. Thank you. If we open the lock, we will find the name of one of the finest gunsmiths of half a century ago—a man whose name has not yet been forgotten—Hamish Graw of Edinburgh. Ah! There! You see?" They did.

Having satisfied himself on that point, Lord Darcy closed the lock again. "Now, men, we have the gun located. We also know that a guest in this very castle is Laird Duncan of Duncan. The Duncan of Duncan himself. A Scot's laird who was, fifteen years ago, His Majesty's Minister Plenipotentiary to the Free Grand Duchy of Milan. That suggests to me that it would be indeed odd if there were not some connection between Laird Duncan and this gun. Eh?"

"Come, come, Master Sean," said Lord Darcy, rather impatiently. "We haven't all the time in the world."

"Patience, my lord; patience," said the little sorcerer calmly. "Can't hurry these things, you know." He was kneeling in front of a large, heavy traveling chest in the bedroom of the guest apartment occupied temporarily by Laird and Lady Duncan, working with the lock. "One position of a lock is just as relevant as the other so you can't work with the bolt. But the pin-tumblers in the cylinder, now, that's a different matter. A lock's built so that the breaks in the tumblers are not related to the surface of the cylinder when the key is out, but there is a relation when the key's *in*, so by taking advantage of that relevancy—Ah!"

The lock clicked open.

Lord Darcy raised the lid gently.

"Carefully, my lord!" Master Sean said in a warning voice. "He's got a spell on the thing! Let me do it." He made Lord Darcy stand back and then lifted the lid of the heavy trunk himself. When it was leaning back against the wall, gaping open widely on its hinges, Master Sean took a long look at the trunk and its lid without touching either of them. There was a second lid on the trunk, a thin one obviously operated by a simple bolt.

Master Sean took his sorcerer's staff, a five-foot, heavy rod made of the wood of the quicken tree or mountain ash, and touched the inner lid. Nothing happened. He touched the bolt. Nothing.

"Hm-m-m," Master Sean murmured thoughtfully. He glanced around the room, and his eyes fell on a heavy stone doorstop. "That ought to do it." He walked over, picked it up, and carried it back to the chest. Then he put it on the rim of the chest in such a position that if the lid were to fall it would be stopped by the doorstop.

Then he put his hand in as if to lift the inner lid.

The heavy outer lid swung forward and down of its own accord, moving with blurring speed, and slammed viciously against the doorstop.

Lord Darcy massaged his right wrist gently, as if he felt where the lid would have hit if he had tried to open the inner lid. "Triggered to slam if a human being sticks a hand in there, eh?"

"Or a head, my lord. Not very effectual if you know what to look for. There are better spells than that for guarding things. Now we'll see what his lordship wants to protect so badly that he practices sorcery without a license." He lifted the lid again, and then opened the inner lid. "It's safe now, my lord. *Look at this!*"

Lord Darcy had already seen. Both men looked in silence at the collection of paraphernalia on the first tray of the chest. Master Sean's busy fingers carefully opened the tissue paper packing of one after another of the objects. "A human skull," he said. "Bottles of graveyard earth. Hm-m-m—this one is labeled 'virgin's blood.' And this! A Hand of Glory!"

It was a mummified human hand, stiff and dry and brown, with the fingers partially curled, as though they were holding an invisible ball three inches or so in diameter. On each of the fingertips was a short candle-stub. When the hand was placed on its back, it would act as a candelabra.

"That pretty much settles it, eh, Master Sean?" Lord Darcy said.

"Indeed, my lord. At the very least, we can get him for possession of materials. Black magic is a matter of symbolism and intent."

"Very well. I want a complete list of the contents of that chest. Be sure to replace everything as it was and relock the trunk." He tugged thoughtfully at an earlobe. "So Laird Duncan has the Talent, eh? Interesting."

"Aye. But not surprising, my lord," said Master Sean without looking up from his work. "It's in the blood. Some attribute it to the Dedannans, who passed through Scotland before they conquered Ireland three thousand years ago, but, however that may be, the Talent runs strong in the Sons of Gael. It makes me boil to see it misused."

While Master Sean talked, Lord Darcy was prowling around the room, reminding one of a lean tomcat who was certain that there was a mouse concealed somewhere.

"It'll make Laird Duncan boil if he isn't stopped," Lord Darcy murmured absently.

"Aye, my lord," said Master Sean. "The mental state necessary to use the Talent for black sorcery is such that it invariably destroys the user—but, if he knows what he's doing, a lot of other people are hurt before he finally gets his."

Lord Darcy opened the jewel box on the dresser. The usual traveling jewelry—enough, but not a great choice.

"A man's mind turns in on itself when he's taken up with hatred and thoughts of revenge," Master Sean droned on. "Or, if he's the type who *enjoys* watching others suffer, or the type who doesn't care but is willing to do anything for gain, then his mind is already warped and the misuse of the Talent just makes it worse."

Lord Darcy found what he was looking for in a drawer, just underneath some neatly folded lingerie. A small holster, beautifully made of Florentine leather, gilded and tooled. He didn't need Master Sean's sorcery to tell him that the little pistol fit it like a hand in a glove.

Father Bright felt as though he had been walking a tightrope for hours. Laird and Lady Duncan had been talking in low, controlled voices that betrayed an inner nervousness, but Father Bright realized that he and the Countess had been doing the same thing. The Duncan of Duncan had offered his condolences on the death of the late Count with the proper air of suppressed sorrow, as had Mary, Lady Duncan. The Countess had accepted them solemnly and with gratitude. But Father Bright was well aware that no one in the room—possibly, he thought, no one in the world—regretted the Count's passing.

Laird Duncan sat in his wheelchair, his sharp Scots features set in a sad smile that showed an intent to be affable even though great sorrow weighed heavily upon him. Father Bright noticed it and realized that his own face had the same sort of expression. No one was fooling anyone else, of that the priest was certain—but for anyone to admit it would be the most boorish breach of etiquette. But there was a haggardness, a look of increased age about the Laird's countenance that Father Bright did not like. His priestly intuition told him clearly that there was a turmoil of emotion in the Scotsman's mind that was ... well, *evil* was the only word for it.

Lady Duncan was, for the most part, silent. In the past fifteen minutes, since she and her husband had come to the informal tea, she had spoken scarcely a dozen words. Her face was masklike, but there was the same look of haggardness about her eyes as there was in her husband's face. But the priest's emphatic sense told him that the emotion here was fear, simple and direct. His keen eyes had noticed that she wore a shade too much make-up. She had almost succeeded in covering up the faint bruise on her right cheek, but not completely.

My lady the Countess D'Evreux was all sadness and unhappiness, but there was neither fear nor evil there. She smiled politely and talked quietly. Father Bright would have been willing to bet that not one of the four of them would remember a word that had been spoken.

Father Bright had placed his chair so that he could keep an eye on the open doorway and the long hall that led in from the Great Keep. He hoped Lord Darcy would hurry. Neither of the guests had been told that the Duke's Investigator was here, and Father Bright was just a little apprehensive about the meeting. The Duncans had not even been told that the Count's death had been murder, but he was certain that they knew.

Father Bright saw Lord Darcy come in through the door at the far end of the hall. He murmured a polite excuse and rose. The other three accepted his excuses with the same politeness and went on with their talk. Father Bright met Lord Darcy in the hall.

"Did you find what you were looking for, Lord Darcy?" the priest asked in a low tone.

"Yes," Lord Darcy said. "I'm afraid we shall have to arrest Laird Duncan."

"Murder?"

"Perhaps. I'm not yet certain of that. But the charge will be black magic. He has all the paraphernalia in a chest in his room. Master Sean reports that a ritual was enacted in the bedroom last night. Of course, that's out of my jurisdiction. You, as a representative of the Church, will have to be the arresting officer." He paused. "You don't seem surprised, Reverence."

"I'm not," Father Bright admitted. "I felt it. You and Master Sean will have to make out a sworn deposition before I can act."

"I understand. Can you do me a favor?"

"If I can."

"Get my lady the Countess out of the room on some pretext or other. Leave me alone with her guests. I do not wish to upset my lady any more than absolutely necessary."

"I think I can do that. Shall we go in together?"

"Why not? But don't mention why I am here. Let them assume I am just another guest."

"Very well."

All three occupants of the room glanced up as Father Bright came in with Lord Darcy. The introductions were made: Lord Darcy humbly begged the pardon of his hostess for his lateness. Father Bright noticed the same sad smile on Lord Darcy's handsome face as the others were wearing.

Lord Darcy helped himself from the buffet table and allowed the Countess to pour him a large cup of hot tea. He mentioned nothing about the recent death. Instead, he turned the conversation toward the wild beauty of Scotland and the excellence of the grouse shooting there.

Father Bright had not sat down again. Instead, he left the room once more. When he returned, he went directly to the Countess and said, in a low, but clearly audible voice: "My lady, Sir Pierre Morlaix has informed me that there are a few matters that require your attention immediately. It will require only a few moments."

My lady the Countess did not hesitate, but made her excuses immediately. "Do finish your tea," she added. "I don't think I shall be long."

Lord Darcy knew the priest would not lie, and he wondered what sort of arrangement had been made with Sir Pierre. Not that it mattered except that Lord Darcy had hoped it would be sufficiently involved for it to keep the Countess busy for at least ten minutes.

The conversation, interrupted but momentarily, returned to grouse.

"I haven't done any shooting since my accident," said Laird Duncan, "but I used to enjoy it immensely. I still have friends up every year for the season."

"What sort of weapon do you prefer for grouse?" Lord Darcy asked.

"A one-inch bore with a modified choke," said the Scot. "I have a pair that I favor. Excellent weapons."

"Of Scottish make?"

"No, no. English. Your London gunsmiths can't be beat for shotguns."

"Oh. I thought perhaps your lordship had had all your guns made in Scotland." As he spoke, he took the little pistol out of his coat pocket and put it carefully on the table.

There was a sudden silence, then Laird Duncan said in an angry voice: "What is this? Where did you get that?"

Lord Darcy glanced at Lady Duncan, who had turned suddenly pale. "Perhaps," he said coolly, "Lady Duncan can tell us."

She shook her head and gasped. For a moment, she had trouble in forming words or finding her voice. Finally: "No. No. I know nothing. Nothing."

But Laird Duncan looked at her oddly.

"You do not deny that it is your gun, my lord?" Lord Darcy asked. "Or your wife's, as the case may be."

"*Where did you get it?*" There was a dangerous quality in the Scotsman's voice. He had once been a powerful man, and Lord Darcy could see his shoulder muscles bunching.

"From the late Count D'Evreux's bedroom."

"What was it doing there?" There was a snarl in the Scot's voice, but Lord Darcy had the feeling that the question was as much directed toward Lady Duncan as it was to himself.

"One of the things it was doing there was shooting Count D'Evreux through the heart."

Lady Duncan slumped forward in a dead faint, overturning her teacup. Laird Duncan made a grab at the gun, ignoring his wife. Lord Darcy's hand snaked out and picked up the weapon before the Scot could touch it. "No, no, my lord," he said mildly. "This is evidence in a murder case. We mustn't tamper with the King's evidence."

He wasn't prepared for what happened next. Laird Duncan roared something obscene in Scots Gaelic, put his hands on the arms of his wheelchair, and, with a great thrust of his powerful arms and shoulders, shoved himself up and forward, toward Lord Darcy, across the table from him. His arms swung up toward Lord Darcy's throat as the momentum of his body carried him toward the investigator.

He might have made it, but the weakness of his legs betrayed him. His waist struck the edge of the massive oaken table, and most of his forward momentum was lost. He collapsed forward, his hands still grasping toward the surprised Englishman. His chin came down hard on the table top. Then he slid back, taking the tablecloth and the china and silverware with him. He lay unmoving on the floor. His wife did not even stir except when the tablecloth tugged at her head.

Lord Darcy had jumped back, overturning his chair. He stood on his feet, looking at the two unconscious forms.

"I don't think there's any permanent damage done to either," said Dr. Pateley an hour later. "Lady Duncan was suffering from shock, of course, but Father Bright brought her round in a hurry. She's a devout woman, I think, even if a sinful one."

"What about Laird Duncan?" Lord Darcy asked.

"Well, that's a different matter. I'm afraid that his back injury was aggravated, and that crack on the chin didn't do him any good. I don't know whether Father Bright can help him or not. Healing takes the co-operation of the patient. I did all I could for him, but I'm just a chirurgeon, not a practitioner of the Healing Art. Father Bright has quite a good reputation in that line, however, and he may be able to do his lordship some good."

Master Sean shook his head dolefully. "His Reverence has the Talent, there's no doubt of that, but now he's pitted against another man who has it—a man whose mind is bent on self-destruction in the long run."

"Well, that's none of my affair," said Dr. Pateley. "I'm just a technician. I'll leave healing up to the Church, where it belongs."

"Master Sean," said Lord Darcy, "there is still a mystery here. We need more evidence. What about the eyes?"

Master Sean blinked. "You mean the picture test, my lord?"

"I do."

"It won't stand up in court, my lord," said the sorcerer.

"I'm aware of that," said Lord Darcy testily.

"Eye test?" Dr. Pateley asked blankly. "I don't believe I understand."

"It's not often used," said Master Sean. "It is a psychic phenomenon that sometimes occurs at the moment of death—especially a violent death. The violent emotional stress causes a sort of backfiring of the mind, if you see what I mean. As a result, the image in the mind of the dying person is returned to the retina. By using the proper sorcery, this image can be developed and the last thing the dead man saw can be brought out.

"But it's a difficult process even under the best of circumstances, and usually the conditions aren't right. In the first place, it doesn't always occur. It never occurs, for instance, when the person is expecting the attack. A man who is killed in a duel, or who is shot after facing the gun for several seconds, has time to adjust to the situation. Also, death must occur almost instantly. If he lingers, even for a few minutes, the effect is lost. And, naturally if the person's eyes are closed at the instant of death, nothing shows up."

"Count D'Evreux's eyes were open," Dr. Pateley said. "They were still open when we found him. How long after death does the image remain?"

"Until the cells of the retina die and lose their identity. Rarely more than twenty-four hours, usually much less."

"It hasn't been twenty-four hours yet," said Lord Darcy, "and there is a chance that the Count was taken completely by surprise."

"I must admit, my lord," Master Sean said thoughtfully, "that the conditions seem favorable. I shall attempt it. But don't put any hopes on it, my lord."

"I shan't. Just do your best, Master Sean. If there is a sorcerer in practice who can do the job, it is you."

"Thank you, my lord. I'll get busy on it right away," said the sorcerer with a subdued glow of pride.

Two hours later, Lord Darcy was striding down the corridor of the Great Hall, Master Sean following up as best he could, his *caorthainn*-wood staff in one hand and his big carpet bag in the other. He had asked Father Bright and the Countess D'Evreux to meet him in one of the smaller guest rooms. But the Countess came to meet him.

"My Lord Darcy," she said, her plain face looking worried and unhappy, "is it true that you suspect Laird and Lady Duncan of this murder? Because, if so, I must—"

"No longer, my lady," Lord Darcy cut her off quickly. "I think we can show that neither is guilty of murder—although, of course, the black magic charge must still be held against Laird Duncan."

"I understand," she said, "but—"

"Please, my lady," Lord Darcy interrupted again, "let me explain everything. Come."

Without another word, she turned and led the way to the room where Father Bright was waiting.

The priest stood waiting, his face showing tenseness.

"Please," said Lord Darcy. "Sit down, both of you. This won't take long. My lady, may Master Sean make use of that table over there?"

"Certainly, my lord," the Countess said softly, "certainly."

"Thank you my lady. Please, please—sit down. This won't take long. Please."

With apparent reluctance, Father Bright and my lady the Countess sat down in two chairs facing Lord Darcy. They paid little attention to what Master Sean O Lochlainn was doing; their eyes were on Lord Darcy.

"Conducting an investigation of this sort is not an easy thing," he began carefully. "Most murder cases could be easily solved by your Chief Man-at-Arms. We find that well-trained county police, in by far the majority of cases, can solve the mystery easily—and in most cases there is very little mystery. But, by His Imperial Majesty's law, the Chief Man-at-Arms must call in a Duke's Investigator if the crime is insoluble or if it involves a member of the aristocracy. For that reason, you were perfectly correct to call His Highness the Duke as soon as murder had been discovered." He leaned back in his chair. "And it has been clear from the first that my lord the late Count was murdered."

Father Bright started to say something, but Lord Darcy cut him off before he could speak. "By 'murder', Reverend Father, I mean that he did not die a natural death—by disease or heart trouble or accident or what-have-you. I should, perhaps, use the word 'homicide'.

"Now the question we have been called upon to answer is simply this: Who was responsible for the homicide?"

The priest and the countess remained silent, looking at Lord Darcy as though he were some sort of divinely inspired oracle.

"As you know ... pardon me, my lady, if I am blunt ... the late Count was somewhat of a playboy. No. I will make that stronger. He was a satyr, a lecher; he was a man with a sexual obsession.

"For such a man, if he indulges in his passions—which the late Count most certainly did—there is usually but one end. Unless he is a man who has a winsome personality—which he did not—there will be someone who will hate him enough to kill him. Such a man inevitably leaves behind him a trail of wronged women and wronged men.

"One such person may kill him.

"One such person did.

"But we must find the person who did and determine the extent of his or her guilt. That is my purpose.

"Now, as to the facts. We know that Edouard has a secret stairway which led directly to his suite. Actually, the secret was poorly kept. There were many women—common and noble—who knew of the existence of that stairway and knew how to enter it. If Edouard left the lower door unlocked, anyone could come up that stairway. He has another lock in the door of his bedroom, so only someone who was invited could come in, even if she ... or he ... could get into the stairway. He was protected.

"Now here is what actually happened that night. I have evidence, by the way, and I have the confessions of both Laird and Lady Duncan. I will explain how I got those confessions in a moment.

"*Primus*: Lady Duncan had an assignation with Count D'Evreux last night. She went up the stairway to his room. She was carrying with her a small pistol. She had had an affair with Edouard, and she had been rebuffed. She was furious. But she went to his room.

"He was drunk when she arrived—in one of the nasty moods with which both of you are familiar. She pleaded with him to accept her again as his mistress. He refused. According to Lady Duncan, he said: 'I don't want you! You're not fit to be in the same room with *her*!'

"The emphasis is Lady Duncan's, not my own.

"Furious, she drew a gun—the little pistol which killed him."

The Countess gasped. "But Mary *couldn't* have—"

"*Please!*" Lord Darcy slammed the palm of his hand on the arm of his chair with an explosive sound. "My lady, you *will* listen to what I have to say!"

He was taking a devil of a chance, he knew. The Countess was his hostess and had every right to exercise her prerogatives. But Lord Darcy was counting on the fact that she had been under Count D'Evreux's influence so long that it would take her a little time to realize that she no longer had to knuckle under to the will of a man who shouted at her. He was right. She became silent.

Father Bright turned to her quickly and said: "Please, my daughter. Wait."

"Your pardon, my lady," Lord Darcy continued smoothly. "I was about to explain to you why I know Lady Duncan could not have killed your brother. There is the matter of the dress. We are certain that the gown that was found in Edouard's closet was worn by the killer. *And that gown could not possibly have fit Lady Duncan!* She's much too ... er ... hefty.

"She has told me her story, and, for reasons I will give you later, I believe it. When she pointed the gun at your brother, she really had no intention of killing him. She had no intention of pulling the trigger. Your brother knew this. He lashed out and slapped the side of her head. She dropped the pistol and fell, sobbing, to the floor. He took her roughly by the arm and 'escorted' her down the stairway. He threw her out.

"Lady Duncan, hysterical, ran to her husband.

"And then, when he had succeeded in calming her down a bit, she realized the position she was in. She knew that Laird Duncan was a violent, a warped man—very similar to Edouard, Count D'Evreux. She dared not tell him the truth, but she had to tell him something. So she lied.

"She told him that Edouard had asked her up in order to tell her something of importance; that that 'something of importance' concerned Laird Duncan's safety; that the Count told her that he knew of Laird Duncan's dabbling in black magic; that he threatened to inform Church authorities on Laird Duncan unless she submitted to his desires; that she had struggled with him and ran away."

Lord Darcy spread his hands. "This was, of course, a tissue of lies. But Laird Duncan believed everything. So great was his ego that he could not believe in her infidelity, although he has been paralyzed for five years."

"How can you be certain that Lady Duncan told the truth?" Father Bright asked warily.

"Aside from the matter of the gown—which Count D'Evreux kept only for women of the common class, *not* the aristocracy—we have the testimony of the actions of Laird Duncan himself. We come then to—

"*Secundus*: Laird Duncan could not have committed the murder physically. *How could a man who was confined to a wheelchair go up that flight of stairs?* I submit to you that it would have been physically impossible.

"The possibility that he has been pretending all these years, and that he is actually capable of walking, was disproved three hours ago, when he actually injured himself by trying to throttle me. His legs are incapable of carrying him even one step—much less carrying him to the top of that stairway."

Lord Darcy folded his hands complacently.

"There remains," said Father Bright, "the possibility that Laird Duncan killed Count D'Evreux by psychical, by magical means."

Lord Darcy nodded. "That is indeed possible, Reverend Sir, as we both know. But not in this instance. Master Sean assures me, and I am certain that you will concur, that a man killed by sorcery, by black magic, dies of internal malfunction, not of a bullet through the heart.

"In effect, the Black Sorcerer induces his enemy to kill himself by psychosomatic means. He dies by what is technically known as psychic induction. Master Sean informs me that the commonest—and crudest—method of doing this is by the so-called 'simalcrum induction' method. That is, by the making of an image—usually, but not necessarily, of wax—and, using the Law of Similarity, inducing death. The Law of Contagion is also used, since the fingernails, hair, spittle, and so on, of the victim are usually incorporated into the image. Am I correct, Father?"

The priest nodded. "Yes. And, contrary to the heresies of certain materialists, it is not at all necessary that the victim be informed of the operation—although, admittedly, it can, in certain circumstances, aid the process."

"Exactly," said Lord Darcy. "But it is well known that material objects can be moved by a competent sorcerer—'black' or 'white'. Would you explain to my lady the Countess why her brother could not have been killed in that manner?"

Father Bright touched his lips with the tip of his tongue and then turned to the girl sitting next to him. "There is a lack of relevancy. In this case, the bullet must have been relevant either to the heart or to the gun. To have traveled with a velocity great enough to penetrate, the relevancy to the heart must have been much greater than the relevancy to the gun. Yet the test, witnessed by myself, that was performed by Master Sean indicates that this was not so. The bullet returned to the gun, not to your brother's heart. The evidence, my dear, is conclusive that the bullet was propelled by purely physical means, and was propelled from the gun."

"Then what was it Laird Duncan did?" the Countess asked.

"*Tertius:*" said Lord Darcy. "Believing what his wife had told him, Laird Duncan flew into a rage. He determined to kill your brother. He used an induction spell. But the spell backfired and almost killed him.

"There are analogies on a material plane. If one adds mineral spirits and air to a fire, the fire will be increased. But if one adds ash, the fire will be put out.

"In a similar manner, if one attacks a living being psychically it will die—but if one attacks a dead thing in such a manner, the psychic energy will be absorbed, to the detriment of the person who has used it.

"In theory, we could charge Laird Duncan with attempted murder, for there is no doubt that he did attempt to kill your brother, my lady. *But your brother was already dead at the time!*

"The resultant dissipation of psychic energy rendered Laird Duncan unconscious for several hours, during which Lady Duncan waited in suspenseful fear.

"Finally, when Laird Duncan regained consciousness, he realized what had happened. He knew that your brother was already dead when he attempted the spell. He thought, therefore, that Lady Duncan had killed the Count.

"On the other hand, Lady Duncan was perfectly well aware that she had left Edouard alive and well. So she thought the black magic of her husband had killed her erstwhile lover."

"Each was trying to protect the other," Father Bright said. "Neither is completely evil, then. There may be something we can do for Laird Duncan."

"I wouldn't know about that, Father," Lord Darcy said. "The Healing Art is the Church's business, not mine." He realized with some amusement that he was paraphrasing Dr. Pateley. "What Laird Duncan had not known," he went on quickly, "was that his wife had taken a gun up to the Count's bedroom. That put a rather different light on her visit, you see. That's why he flew into such a towering rage at me—not because I was accusing him or his wife of murder, but because I had cast doubt on his wife's behavior."

He turned his head to look at the table where the Irish sorcerer was working. "Ready, Master Sean?"

"Aye, my lord. All I have to do is set up the screen and light the lantern in the projector."

"Go ahead, then." He looked back at Father Bright and the Countess. "Master Sean has a rather interesting lantern slide I want you to look at."

"The most successful development I've ever made, if I may say so, my lord," the sorcerer said.

"Proceed."

Master Sean opened the shutter on the projector, and a picture sprang into being on the screen.

There were gasps from Father Bright and the Countess.

It was a woman. She was wearing the gown that had hung in the Count's closet. A button had been torn off, and the gown gaped open. Her right hand was almost completely obscured by a dense cloud of smoke. Obviously she had just fired a pistol directly at the onlooker.

But that was not what had caused the gasps.

The girl was beautiful. Gloriously, ravishingly beautiful. It was not a delicate beauty. There was nothing flower-like or peaceful in it. It was a beauty that could have but one effect on a normal human male. She was the most physically desirable woman one could imagine.

Retro mea, Sathanas, Father Bright thought wryly. *She's almost obscenely beautiful.*

Only the Countess was unaffected by the desirability of the image. She saw only the startling beauty.

"Has neither of you seen that woman before? I thought not," said Lord Darcy. "Nor had Laird or Lady Duncan. Nor Sir Pierre.

"Who is she? We don't know. But we can make a few deductions. She must have come to the Count's room by appointment. This is quite obviously the woman Edouard mentioned to Lady Duncan—the woman, the 'she' that the Scots noblewoman could not compare with. It is almost certain she is a commoner; otherwise she would not be wearing a robe from the Count's collection. She must have changed right there in the bedroom. Then she and the Count quarreled—about what, we do not know. The Count had previously taken Lady Duncan's pistol away from her and had evidently carelessly let it lay on that table you see behind the girl. She grabbed it and shot him. Then she changed clothes again, hung up the robe, and ran away. No one saw her come or go. The Count had designed the stairway for just that purpose.

"Oh, we'll find her, never fear—now that we know what she looks like.

"At any rate," Lord Darcy concluded, "the mystery is now solved to my complete satisfaction, and I shall so report to His Highness."

Richard, Duke of Normandy, poured two liberal portions of excellent brandy into a pair of crystal goblets. There was a smile of satisfaction on his youthful face as he handed one of the goblets to Lord Darcy. "Very well done, my lord," he said. "Very well done."

"I am gratified to hear Your Highness say so," said Lord Darcy, accepting the brandy.

"But how were you so certain that it was *not* someone from outside the castle? Anyone could have come in through the main gate. That's always open."

"True, Your Highness. But the door at the foot of the stairway was *locked*. Count D'Evreux locked it after he threw Lady Duncan out. There is no way of locking or unlocking it from the outside; the door had not been forced. No one could have come in that way, nor left that way, after Lady Duncan was so forcibly ejected. The only other way into the Count's suite was by the other door, and that door was unlocked."

"I see," said Duke Richard. "I wonder why she went up there in the first place?"

"Probably because he asked her to. Any other woman would have known what she was getting into if she accepted an invitation to Count D'Evreux's suite."

The Duke's handsome face darkened. "No. One would hardly expect that sort of thing from one's own brother. She was perfectly justified in shooting him."

"Perfectly, Your Highness. And had she been anyone but the heiress, she would undoubtedly have confessed immediately. Indeed, it was all I could do to keep her from confessing to me when she thought I was going to charge the Duncans with the killing. But she knew that it was necessary to preserve the reputation of her brother and herself. Not as private persons, but as Count and Countess, as officers of the Government of His Imperial Majesty the King. For a man to be known as a rake is one thing. Most people don't care about that sort of thing in a public official so long as he does his duty and does it well—which, as Your Highness knows, the Count did.

"But to be shot to death while attempting to assault his own sister—that is quite another thing. She was perfectly justified in attempting to cover it up. And she will remain silent unless someone else is accused of the crime."

"Which, of course, will not happen," said Duke Richard. He sipped at the brandy, then said: "She will make a good Countess. She has judgment and she can keep cool under duress. After she had shot her own brother, she might have panicked, but she didn't. How many women would have thought of simply taking off the damaged gown and putting on its duplicate from the closet?"

"Very few," Lord Darcy agreed. "That's why I never mentioned that I knew the Count's wardrobe contained dresses identical to her own. By the way, Your Highness, if any good Healer, like Father Bright, had known of those duplicate dresses, he would have realized that the Count had a sexual obsession about his sister. He would have known that all the other women the Count went after were sister substitutes."

"Yes; of course. And none of them measure up." He put his goblet on the table. "I shall inform the King my brother that I recommended the new Countess whole-heartedly. No word of this must be put down in writing, of course. You know and I know and the King must know. No one else must know."

"One other knows," said Lord Darcy.

"Who?" The Duke looked startled.

"Father Bright."

Duke Richard looked relieved. "Naturally. He won't tell her that we know, will he?"

"I think Father Bright's discretion can be relied upon."

In the dimness of the confessional, Alice, Countess D'Evreux knelt and listened to the voice of Father Bright.

"I shall not give you any penance, my child, for you have committed no sin—that is, in so far as the death of your brother is concerned. For the rest of your sins, you must read and memorize the third chapter of 'The Soul and The World,' by St. James Huntington."

He started to pronounce the absolution, but the Countess said:

"I don't understand one thing. That picture. That wasn't me. I never saw such a gorgeously beautiful girl in my life. And I'm so plain. I don't understand."

"Had you looked more closely, my child, you would have seen that the face did look like yours—only it was idealized. When a subjective reality is made objective, distortions invariably show up; that is why such things cannot be accepted as evidence of objective reality in court." He paused. "To put it another way, my child: Beauty is in the eye of the beholder."

www.ingramcontent.com/pod-product-compliance
Lightning Source LLC
LaVergne TN
LVHW041802190726
843493LV00008B/2751